No Metal No Magic

Element 2

Helium, Presented By Hetha

From The Magical Elements of the Periodic Table Series

Hetha

2	4.003
He	
helium	

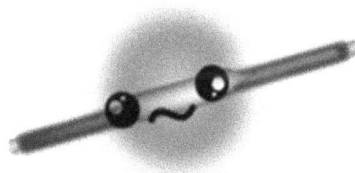

Helium

By Sybrina Durant with Illustrations by Pranavva et al.

No Metal No Magic

Element 2

Helium, Presented By Hetha
From The Magical Elements of the Periodic Table Series

Story copyright 2025

Soft Cover Print ISBN: 978-1-942740-56-8

BISAC Codes:

JNF051070 JUVENILE NONFICTION / Science & Nature / Chemistry

JNF016000 JUVENILE NONFICTION / Curiosities & Wonders

JNF051080 JUVENILE NONFICTION / Science & Nature / Earth Sciences / General

Soft Cover Print ISBN 13 - 978-1-942740-56-8

Hetha Presents Helium

This Element 2 book features the periodic table element, Helium. It is presented by Hetha, an Alchemical Wizard who wields a magical elemental staff with powers based on its periodic table element.

Hetha is just one of the 118 elementals who will present all of the Magical Elements of the Periodic Table to readers who are curious about the wonders of the world.

Hetha introduces the very airy element, Helium, in her book.

The Alchemical Wizards and their other techno-magical friends are the perfect group to introduce you to the elements in the Periodic Table. Hopefully, this Magical Elements of the periodic table book will spark an interest in the magical and real world properties of all the elements known today. You may be surprised at how prominently they feature in our every day lives.

Each page in this book contains terms that might not be completely familiar to the reader. Refer to the definitions in the back of the book to get a clear understanding of each meaning.

There is also a fun elemental themed Periodic Table at the back of the book. It features 118 elements presented by fanciful characters like unicorns, dragons, wizards, knights and goblins.. They want you to remember that if there's no metal...there's no magic or technology.

Remember, "No metal – No Magic. . .and No Technology".

It's Techo-Magical.

Note: Sybrina Publishing websites are Sybrina.com and MagicalPTElements.com. Follow sybrinapublishing on Instagram, Magical Elements of the Periodic Table on Facebook, @sybrinad on Pinterest, Sybrina_SPT on Twitter; and Sybrina Durant on LinkedIn.

Hetha

The Wizard With The Helium Staff

Symbol: He Atomic Number: 2 Atomic Mass: 4.003

Helium resides in Group 18 Period 1 on the Periodic Table.

The atomic symbol is He. It's Atomic Number is 2. It's Atomic Mass is 4.003.

Magical elementals from the Magical Elements of the Periodic Table books present all of the elements of the periodic table in fantastical and real life terms.

In the books, each elemental character has magical powers based on the properties of the elements that come from the land, air and water. They are the perfect group to introduce you to metals, metalloids, non-metals, halogens, noble gases and much more.

Unicorns, dragons, alchemists, knights, and goblins will show you how people of this world always have and always will depend upon the elements that our earth provides for all of our needs.

Use this Periodic Table as you would any other to spark an interest in the magical and real world properties of all the elements known today. You may be surprised at how prominently they feature in our every day lives.

Magical Elements of The Periodic Table

No Metal

Actinium To Zirconium

No Magic

Remember, "No Metal—No Magic." . . .And no technology.

It's Techno-Magical

LEGEND

Alkali Metals
Alkali Earth Metals
Transition Metals
Post-Transition (or Other Metals)
Metalloids
Non-Metals
Halogens
Noble Gases
Rare Earth Lanthanide Metals
Actinide Metals
Super Heavy—Radioactive

Alloys are created when 2 or more metals are combined. Compounds are created when 2 or more non-metals are combined.

EXAMPLE OF A COMPOUND

Quincy

Quick Lime = Ca + O

Used for Concrete

Both Carbon and Oxygen are reactive nonmetals.

EXAMPLE OF AN ALLOY

White Wing

Used for jewelry, dental amalgams plus connectors, and switch and relay contacts for electronics.

White Gold = (White) Ni, Zn, Au, Cu, Ag (Gold)

Includes 58.5 % gold, 22% copper, 8% zinc, 7% nickel, 4.5% silver and possibly other elements.

Sybrina.com

He — helium

2   4.003

He

helium

Hetha

Balloons

Helium is a Noble Gas

Helium was first observed in 1868 by French astronomer Pierre-Jules-César Janssen during a solar eclipse in Guntur, India. In 1895, Scottish chemist William Ramsay finally isolated helium from the mineral cleveite in London.

Helium is a colorless, odorless, and tasteless gas at room temperature. When electrified, it glows pink-orange.

Helium is the least reactive of all elements; it does not react with any other elements or ions, so there are no helium-bearing minerals in nature.

Helium has high thermal conductivity meaning it can conduct heat very quickly and effectively.

Helium is a Monoatomic Noble Gas. Monoatomic means it exists as individual atoms rather than molecules formed by two or more atoms bonded together.

LEGEND

Alkali Metals
Alkali Earth Metals
Transition Metals
Post-Transition (or Other Metals)
Metalloids
Non-Metals
Halogens
Noble Gases
Rare Earth Lanthanide Metals
Actinide Metals
Super Heavy—Radioactive

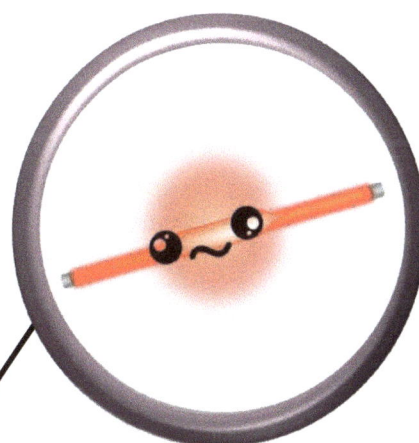

Helium Element

Atomic Structure

He

Noble Gases—These elements reside in column 8. They are all odorless, colorless gases that are chemically very stable (inert). They don't generally form compounds by bonding with another element. These include helium, neon, argon, krypton, xenon, and radon.

Time to enter the cool world of helium, that light-as-a-feather gas that's more than just a party trick for floating balloons. You might be surprised to learn that helium isn't just about making your birthday bash a blast; it's one of the most abundant elements in the universe, and its unique properties make it a magical element in many fields.

First off, helium is a noble gas, meaning it doesn't react with other elements, so it's super safe—unlike hydrogen, which goes "boom!" if it gets too close to a spark. That's why helium is the go-to choice for birthday balloons and giant blimps that float gracefully through the sky. Helium's stability not only ensures a carefree experience for flying things but also makes it invaluable in scientific applications, such as cryogenics and MRI machines. In contrast, the flammability of hydrogen poses significant risks, reinforcing helium's reputation as the safer, more reliable option in various industries and everyday uses.

In the healthcare world, it's a lifesaver (literally). It's mixed with oxygen to help people with breathing issues like asthma. Imagine being able to breathe easier with just a little help from this gas—it's like having a breathing buddy that's always there for you. For those struggling to catch their breath, this simple combination can make the difference between gasping for air and taking a deep, satisfying breath. It's a crucial tool that empowers individuals to reclaim their lives, allowing them to engage in daily activities with renewed confidence and ease.

If you're into diving, then you should know that helium is part of the secret sauce for deep-sea explorers. When mixed with oxygen, it prevents nasty stuff called nitrogen narcosis, which can make divers feel woozy at great depths. So, the next time you see a diver, remember that helium's got their back. This clever mixture allows divers to manage their mental clarity and physical endurance, enabling them to explore underwater worlds with confidence. After all, a clear head is just as crucial as the right gear when navigating the mysterious depths of the ocean.

In the world of technology, Helium plays a crucial role in MRI machines, helping doctors see detailed images of our insides. It keeps those super magnets cool so they give us accurate diagnoses. Without helium's unique properties, the magnets would overheat, compromising the clarity of the images. Thus, this noble gas becomes essential in the pursuit of precise and effective medical imaging. Speaking of keeping things cool, helium is essential in particle accelerators and advanced scientific tools that require ultra-low temperatures.

Helium plays a crucial role in the manufacturing of semiconductors for our smartphones and keeps hard drives cool. Its unique properties not only enhance the efficiency of electronic devices but also extend their longevity, ensuring we enjoy seamless performance in our daily technology.

In aerospace, it's used to pressurize fuel tanks, ensuring rockets don't go kaboom during launches. This vital process not only prevents catastrophic failures but also helps maintain optimal fuel flow and performance during the flight. By carefully managing the pressure within these tanks, engineers can maximize efficiency and enhance the safety of daring space missions.

Let's not forget about safety—helium is key in deploying airbags in cars, making sure everyone stays safe during a crash. It's like an unsung hero that works behind the scenes to protect us when it matters most.

With all these amazing uses, helium is in high demand, and keeping it flowing is super important for tech, healthcare, and even fun festivities! So, the next time you see a balloon floating in the air, remember that there's a lot more to helium than meets the eye. It's a small gas with massive impact.

Uses For Helium

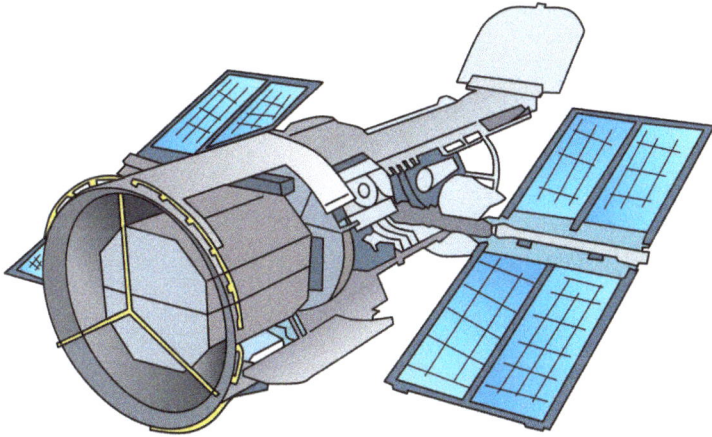

Helium is used in cooling gases for space telescopes, enabling them to operate at extremely low temperatures essential for capturing faint astronomical signals. This remarkable property of helium not only enhances the performance of these instruments but also allows scientists to capture detailed images of distant celestial bodies that would otherwise remain obscured.

It is used as a lifting gas for urban air mobility vehicles and drones. This lightweight gas enhances the efficiency and maneuverability of these innovative crafts, making them ideal for urban environments. As cities evolve and adapt to new technologies, the utilization of such gases can significantly reduce energy consumption and emissions, paving the way for a more sustainable future in urban air transit.

Uses For Helium

(Continued)

Helium is mixed with oxygen and nitrogen in scuba gear to create breathing mixtures that prevent nitrogen narcosis (extreme sleepiness). This careful blend allows divers to maintain cognitive function and clarity while exploring the depths of the ocean. Additionally, using helium reduces the risk of gas embolisms, making deep dives safer and more enjoyable for underwater adventurers.

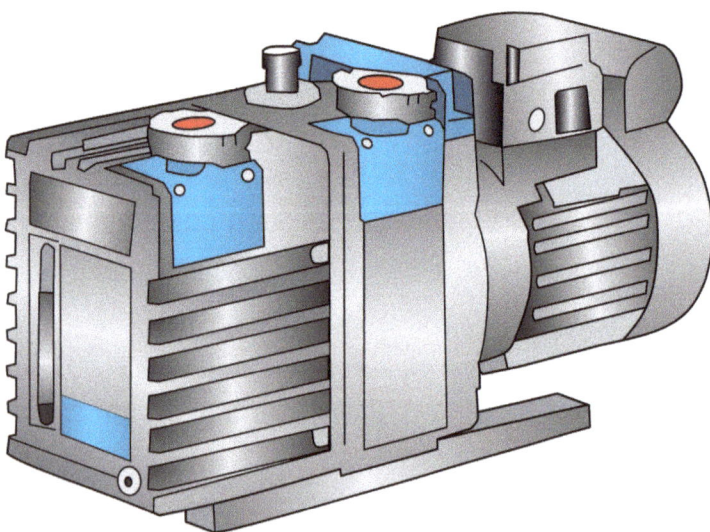

Helium is used in the operation of high-performance vacuum pumps. Its unique low viscosity allows for efficient gas flow and rapid evacuation of air, making it essential in achieving the deep vacuum levels required for various industrial processes. Additionally, helium's inert properties prevent contamination, ensuring the integrity of sensitive systems during operation.

Uses For Helium

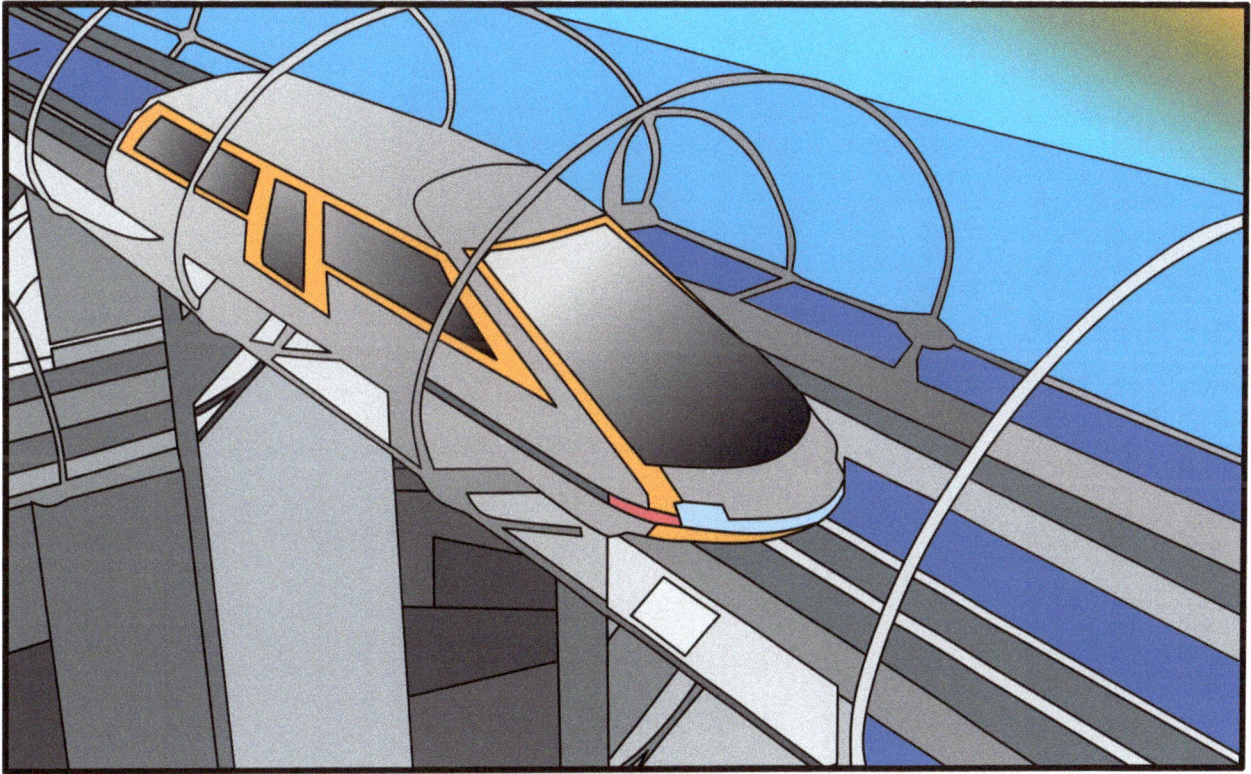

Helium could be used in the cooling and pressure systems of hyperloop transport technologies. Its unique properties make it an ideal candidate for maintaining optimal temperatures, thereby enhancing the efficiency of the system. Additionally, being non-reactive and lightweight, helium can contribute to the overall safety and performance of the hyperloop, ensuring a smooth and reliable travel experience for passengers.

The Source of Helium

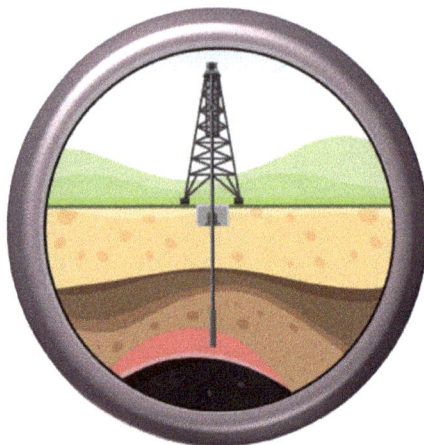

Helium is found deep under the Earth's surface along with other gases. It's mostly taken from natural gas. The U.S., Qatar, and Algeria have the biggest helium reserves in the world.

Helium is a fascinating gas that undergoes a lengthy and complex journey before it ends up in party balloons or serves as a cooling element in scientific research. The story of helium begins deep beneath the Earth's surface, primarily found in natural gas reserves. However, there's an interesting twist: helium exists in only minute quantities. For extraction to be worthwhile, the gas must be located in concentrations of at least 0.3%. To access this elusive gas, scientists begin by drilling into these natural gas deposits, which involves making deep holes in the ground.

When the drilling is successful, natural gas, which is primarily made up of methane, begins to flow up through pipes designed to transport it to processing plants. Upon arrival, this gas is like discovering a hidden treasure. At this stage, a "crude helium stream" can be extracted that contains about 50-70% helium mixed with other gases. However, this mixture is just a preliminary step; it isn't yet pure helium. Much like uncovering a trove of valuable gems, the true work begins here.

The Source of Helium (Continued)

Once this gas has been adequately purified, it is then prepared for storage and distribution. The purified helium is compressed into high-pressure tanks or cylinders, enabling it to be safely transported to various locations. This step is crucial, as helium is in high demand not only for filling balloons at celebrations, but also for critical applications in scientific research and medical technology.

Helium is regularly used in laboratories for various purposes, ranging from cooling superconducting magnets in MRI machines to creating controlled environments for specific experiments. Its low boiling point and inert nature make it an ideal choice for these applications. The journey of helium from the earth to our everyday lives underscores the importance of this unique gas and the sophisticated procedures required to make it available.

But the story doesn't end with extraction and purification. The use of helium has become a significant part of industries and applications beyond what many realize. For instance, manufacturers used helium during the production of fiber optics and semiconductors, demonstrating its integral role in modern technology. Furthermore, it is essential in the aerospace sector for filling airships and providing an inert atmosphere for certain rocket propellants.

The quest for helium doesn't come without challenges. While it is abundant in the universe—particularly in stars where it is formed through nuclear fusion—on Earth, its scarcity and the complexities involved in extraction have made it a valuable resource. Additionally, political and economic factors complicate the helium market, as countries that store large helium reserves sometimes face pressures to maintain a balance between supply and demand.

Moreover, there is an increasing awareness of helium conservation due to its non-renewable nature. It is noteworthy to mention that once helium is released into the atmosphere, it eventually escapes into space, making it a resource that cannot be replenished. This reality raises concerns about sustainable practices surrounding helium extraction and usage, prompting discussions about whether we should limit its use in non-essential applications, such as party balloons.

The journey of helium from deep within the Earth to its myriad uses highlights a remarkable process that combines geology, engineering, and chemistry. From its discovery in natural gas reserves to the intricate methods of purification and storage, helium indeed has a story worth telling. Its critical roles in scientific advancements and everyday life remind us of the importance of managing this precious resource wisely. As we continue to enjoy the benefits of helium, it becomes vital to recognize both its significance and the responsibility we share in preserving it for future generations.

Magical Elements of The Periodic Table

Magical elementals from the Magical Elements of the Periodic Table books present all of the elements of the periodic table in fantastical and real life terms.

In the books, each elemental character has magical powers based on the properties of the elements that come from the land, air and water. They are the perfect group to introduce you to metals, metalloids, non-metals, halogens, noble gases and much more.

Unicorns, dragons, alchemists, knights, and goblins will show you how people of this world always have and always will depend upon the elements that our earth provides for all of our needs.

Use this Periodic Table as you would any other to spark an interest in the magical and real world properties of all the elements known today. You may be surprised at how prominently they feature in our every day lives.

No Metal

No Magic

Actinium To Zirconium

Remember, "No Metal—No Magic." . . .And no technology.

Group	Element	Character	Use
1 H 1.008 hydrogen	Hilay	Textile Manufacturing
3 Li 6.94 lithium	Lillian	Batteries
11 Na 22.99 sodium	Sorn	Salt
19 K 39.10 potassium	Pearl	Saline Drips
37 Rb 85.47 rubidium	Ruby	Night Vision Glasses
2 He 4.003 helium	Hetha	Balloons
4 Be 9.012 beryllium	Berwyn	Musical Instrument
12 Mg 24.31 magnesium	Maggie	In Your Bones
20 Ca 40.08 calcium	Verly	Teeth
38 Sr 87.62 strontium	Strauna	Computer Screens
21 Sc 44.96 scandium	Sandra	Bicycles
39 Y 88.91 yttrium	Yago	Microwave
22 Ti 47.87 titanium	Tilly	Aerospace
40 Zr 91.22 zirconium	Zora	Chemical Pipelines
23 V 50.94 vanadium	Vana	Black Printer Ink
41 Nb 92.91 niobium	Nomah	Mag Lev Trains
24 Cr 52.00 chromium	Crowmist	Stainless Steel
42 Mo 95.95 molybdenum	Maximo	Cutting Tools
25 Mn 54.94 manganese	Mangar	Earth Movers
43 Tc 98 technetium	Tephen	Radio Active Diagnosis
26 Fe 55.85 iron	Iown	Bicycle Chains
44 Ru 101.1 ruthenium	Ruth	Electrical Switches
27 Co 58.93 cobalt	Coriss	Magnets
45 Rh 102.9 rhodium	Rovana	Finish for Jewelry
28 Ni 58.69 nickel	Nix	Guitar Strings
46 Pd 106.4 palladium	Paediln	Concert Flute
29 Cu 63.55 copper	Cuprum	Money
47 Ag 107.9 silver	Silubhra	Ventilator
30 Zn 65.38 zinc	Dr. Zinko	Suntan Lotion
48 Cd 112.4 cadmium	Cadmus	Power Tools
5 B 10.81 boron	Boroleas	Sports Equipment
13 Al 26.98 aluminium	Alunna	Airplanes
31 Ga 69.72 gallium	Gallant	LED Displays
49 In 114.8 indium	Iker	Liquid Crystal Display (LCD)
6 C 12.01 carbon	Cole	Charcoal
14 Si 28.09 silicon	Silonar	Glass
32 Ge 72.63 germanium	Gemel	Camera Lense
50 Sn 118.7 tin	Tinam	Liquid Crystal Display
7 N 14.01 nitrogen	Nitra	Food Packaging
15 P 30.97 phosphorus	Phova	Fertilizer
33 As 74.92 arsenic	Arlyn	Poison
51 Sb 121.8 antimony	Antz	Flame Resistant Fabric
8 O 16.00 oxygen	Ozzy	Air
16 S 32.06 sulfur	Xoe	Matches
34 Se 78.97 selenium	Selenice	Printers
52 Te 127.6 tellurium	Tellan	Vulcanized Rubber
9 F 19.00 fluorine	Fleure	Strong Bones and Teeth
17 Cl 35.45 chlorine	Krystl	Swimming Pools
35 Br 79.90 bromine	Bropeh	Photography Film
53 I 126.9 iodine	Jody	Cloud Seeding
10 Ne 20.18 neon	Jalan	Advertising Signs
18 Ar 39.95 argon	Areg	Light Bulbs
36 Kr 83.80 krypton	Krypto	Detect Leaks
54 Xe 131.3 xenon	Xena	Used To Catch Species

It's Techno-Magical

Period 6

Cs caesium — Caelloth — Atomic Clocks	Ba barium — Barsana — Spark Plugs
Hf hafnium — Hallam — Nuclear Submarines	Ta tantalum — Taltra — Mobile Phones
W tungsten (Wolfram) — Wolfie — 3D Printing Nozzles	Re rhenium — Rankin — Rocket Engines
Os osmium — Osm — For Lab Testing	Ir iridium — Iridna — Weight Scale
Pt platinum — Paedra — Pacemaker	Au Gold — Ghel — Pacemakers and Stents
Hg Mercury — Questa — Barometer	Tl thallium — Thanolen — Tattoo Ink
Pb lead — Lauda — Batteries	Bi bismuth — Bitsy — Fire Sprinklers
Po polonium — Polgah — Anti-Static Brushes	At astatine — Aszrad — Thyroid Cancer Treatment
Rn radon — Raman — Earthquake Prediction	

SUPER HEAVY METALS—RADIOACTIVE (Period 7)

87 223 Fr francium — Francine — Cancer Treatment	88 226 Ra radium — Raele — Luminous Watches
104 267 Rf rutherfordium — Rukolz — Radioactive	105 268 Db dubnium — Dubnic — Radioactive
106 269 Sg seaborgium — Starx — Radioactive	107 270 Bh bohrium — Bisadak — Radioactive
108 277 Hs hassium — Hoiga — Radioactive	109 276 Mt meitnerium — Mohdort — Radioactive
110 281 Ds darmstadtium — Dardank — Radioactive	111 282 Rg roentgenium — Rogmort — Radioactive
112 285 Cn copernicium — Clawotz — Radioactive	113 286 Nh nihonium — Nirtak — Radioactive
114 289 Fl flerovium — Fleth — Radioactive	115 290 Mc moscovium — Molit — Radioactive
116 293 Lv livermorium — Ligee — Radioactive	117 294 Ts tennessine — Tubnuk — Radioactive
118 294 Og oganesson — Otyrt — Radioactive	

RARE EARTH LANTHANIDE METALS (57 thru 71)

57 138.9 La lanthanum — Lannion — Telescope Lense	58 140.1 Ce cerium — Cerelia — Lighter Flints
59 140.9 Pr praseodymium — Praetoc — Welder Mask	60 144.2 Nd neodymium — Neluehta — Electric Car Motors
61 145 Pm promethium — Prida — Night Light	62 150.4 Sm samarium — Samrida — Electric Guitar Pickup
63 152.0 Eu europium — Euell — Fluorescent Light	64 157.3 Gd gadolinium — Galoa — MRI Diagnosis
65 158.9 Tb terbium — Torin — Solid State Device	66 162.5 Dy dysprosium — Dypsie — Sonar Sensors
67 164.9 Ho holmium — Holmia — Eye Laser	68 167.3 Er erbium — Erbie — Optical Communications
69 168.9 Tm thulium — Thurwin — Eye Laser	70 173.0 Yb ytterbium — Yitzy — Amplifier Fiber Optics
71 175.0 Lu lutetium — Umi — Positron Emission Tomography (PET)	

ACTINIDE METALS (89 thru 103)

89 227 Ac actinium — Azanas — Radioactive Medicine	90 232.0 Th thorium — Thorfin — Heat Resistant Paint
91 231.0 Pa protactinium — Profit — Radioactive Waste	92 238.0 U uranium — Uri — Used To Power Submarines
93 237 Np neptunium — Napthas — Nuclear Fuel	94 244 Pu plutonium — Pluxtan — Power Satellites
95 243 Am americium — Amerite — Smoke Detector	96 247 Cm curium — Curran — Moon Rover
97 247 Bk berkelium — Berenere — Scientific Research	98 251 Cf californium — Calastian — Metal Detector
99 252 Es einsteinium — Elizama — Nuclear Research	100 257 Fm fermium — Ferley — Scientific Research
101 258 Md mendelevium — Menesant — Scientific Research	102 259 No nobelium — Norlium — Nuclear Research
103 266 Lr lawrencium — Larelis — Radioactive Research	

EXAMPLE OF AN ALLOY

White Wing

White Gold

Includes 58.5% gold, 22% copper, 8% zinc, 7% nickel, 4.5% silver and possibly other elements.

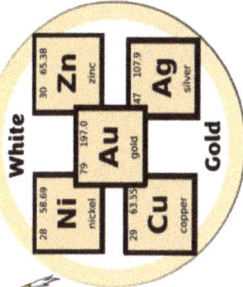

Used for jewelry, dental amalgams plus connectors, and switch and relay contacts for electronics.

28 58.69 Ni nickel	30 65.38 Zn zinc — White
79 197.0 Au gold	47 107.9 Ag silver — Gold
29 63.55 Cu copper	

Alloys are created when 2 or more metals are combined. Compounds are created when 2 or more non-metals are combined.

EXAMPLE OF A COMPOUND

Quincy

Quick Lime =

20 40.08 Ca calcium — Verty — Teeth

+

8 16.00 O oxygen — Ozzy — Air

Used for Concrete

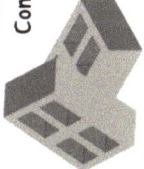

Both Carbon and Oxygen are reactive nonmetals.

Sybrina.com

LEGEND

- Alkali Metals
- Alkali Earth Metals
- Transition Metals
- Post-Transition (or Other Metals)
- Metalloids
- Non-Metals
- Halogens
- Noble Gases
- Rare Earth Lanthanide Metals
- Actinide Metals
- Super Heavy—Radioactive

All Of The Periodic Table Elements Listed Alphabetically

Element Listed In Red Is Featured In This Book

ACTINIUM—AC—89

ALUMINUM—AL—13

AMERICIUM—AM—95

ANTIMONY—SB—51

ARGON—AR—18

ARSENIC—AS—33

ASTATINE—AT—85

BARIUM—BA—56

BERKELIUM—BK—97

BERYLLIUM—BE—4

BISMUTH—BI—83

BOHRIUM—BH—107

BORON—B—5

BROMINE—BR—35

CADMIUM—CD—48

CALCIUM (Vital)—CA—20

CALIFORNIUM—CF—98

CARBON—C—6

CERIUM—CE—58

CESIUM—CS—55

CHLORINE (Keen)—CL—17

CHROMIUM—CR—24

COBALT—CO—27

COPERNICIUM—CN—112

COPPER—CU—29

CURIUM—CM—96

DARMSTADTIUM—DS—110

DUBNIUM—DB—105

DYSPROSIUM—DY—66

ERBIUM—ER—68

EINSTEINIUM—ES—99

EUROPIUM—EU—63

FERMIUM—FM—100

FLEROVIUM—FL—114

FLUORINE—F—9

FRANCIUM—FR—87

GADOLINIUM—GD—64

GALLIUM—GA—31

GERMANIUM—GE—32

GOLD—AU—79

HAFNIUM—HF—72

HASSIUM—HS—108

HELIUM—HE—2

HOLMIUM—HO—67

HYDROGEN—H—1

INDIUM—IN—49

IODINE (JODIUM)—I—53

IRIDIUM—IR—77

IRON—FE—26

KRYPTON—KR—36

LANTHANUM—LA—57

LAWRENCIUM—LR—103

LEAD—PB—82

LITHIUM—LI—3

LIVERMORIUM—LV—116

LUTETIUM (Unique)—LU—71

MAGNESIUM—MG—12

MANGANESE—MN—25

MEITNERIUM—MT—109

MENDELEVIUM—MD—101

MERCURY (QUICK SILVER)—HG—80

MOLYBDENUM—MO—42

MOSCOVIUM—MC—115

NEODYMIUM—ND—60

NEON (Jazzy)—NE—10

NEPTUNIUM—NP—93

NICKEL—NI—28

NIHONIUM—NH—113

NIOBIUM—NB—41

NITROGEN—N—7

NOBELIUM—NO—102

OGANESSON—OG—118

OSMIUM—OS—76

OXYGEN—O—8

PALLADIUM—PD—46

PHOSPHORUS—P—15

PLATINUM—PT—78

PLUTONIUM—PU—94

POLONIUM—PO—84

POTASSIUM—K—19

PRASEODYMIUM—PR—59

PROMETHIUM—PM—61

PROTACTINIUM—PA—91

RADIUM—RA—88

RADON—RN—86

RHENIUM—RE—75

RHODIUM—RH—45

ROENTGENIUM—RG—111

RUBIDIUM—RB—37

RUTHENIUM—RU—44

RUTHERFORDIUM—RF—104

SAMARIUM—SM—62

SCANDIUM—SC—21

SEABORGIUM—SG—106

SELENIUM—SE—34

SILICON—SI—14

SILVER—AG—47

SODIUM—NA—11

STRONTIUM—SR—38

SULFUR (Xanthous)—S—16

TANTALUM—TA—73

TECHNETIUM—TC—43

TELLURIUM—TE—52

TENNESSINE—TS—117

TERBIUM—TB—65

THALLIUM—TI—81

THORIUM—TH—90

THULIUM—TH—69

TIN—SN—50

TITANIUM—TI—22

TUNGSTEN—W (WOLFRAM)—74

URANIUM—U—92

VANADIUM—V—23

XENON—XE—54

YTTERBIUM—YB—70

YTTRIUM—Y—39

ZINC—ZN—30

ZIRCONIUM—ZR—40

Hetha
Presents
Helium

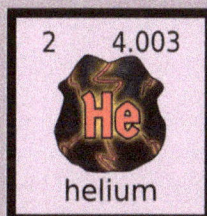

2	4.003
He	
helium	

Did You Know?

The name comes from the Greek word "helios," which means sun, because helium was first found in the sun's corona. French astronomer, Pierre-Jules-César Janssen, discovered helium way back in 1868 when he was checking out a solar eclipse. The sun pumps out an incredible 700 million tons of helium every second. Helium atoms are super light, which is why they can easily float away from Earth's gravity.

Magical Elements of The Periodic Table

- The name, helium, comes from the Greek word "helios," which means sun, because helium was first found in the sun's corona. This fascinating element was discovered on the sun, long before it was found on Earth, making it a cosmic treasure that connects us to the stars.
- The sun pumps out an incredible 700 million tons of helium every second.
- Helium was first found on the sun by a French astronomer named Pierre-Jules-César Janssen way back in 1868 when he was checking out a solar eclipse through a spectroscope.
- Non-reactive gases such as helium, neon, argon, krypton, xenon, and radon do not readily form compounds with other elements and are therefore considered "noble." These colorless, odorless monoatomic elements exist as single atoms rather than as molecules. This means they won't mix or bond with other elements. Because of their stability and lack of reactivity, noble gases are often used in fancy lighting like neon signs and even in lasers, bringing a colorful glow to our nights.

Did You Know? (continued)

- Helium's thermal conductivity (ability to conduct heat) is greater than that of any gas except hydrogen. This means that helium is super efficient at transferring heat, which is why it's often used in applications like cooling superconducting magnets or even in the space industry.

- Helium can transfer heat incredibly efficiently, making it perfect for use in cryogenics, where extreme low temperatures are required. Interestingly, this property is one of the reasons why helium is used in party balloons—beyond its lighter-than-air qualities, it helps keep the air inside the balloon cool and stable for a longer time.

- Helium was once commonly used to create cartoon character voices like Mickey Mouse and Alvin and the Chipmunks. Breathing it in caused the vocal cords of the voice actors to vibrate faster, creating a squeaky, cartoonish sound that perfectly fit those beloved characters. It is no longer used in television or movies because too much can lead to serious health issues.

- Dmitrii Mendeleev, the creator of the periodic table did not believe helium existed until 1895, when English chemist William Ramsay isolated helium from a uranium-rich mineral and then immediately discovered Krypton, Neon, and Xenon. Those other noble gas discoveries finally convinced Mendeleev to add another column to his periodic table.

- Helium atoms are super light, which is why they can easily float away from Earth's gravity. Once helium escapes into the air, this light gas is gone forever. This super-light gas is so buoyant that Earth's gravity can't keep it tethered. The helium that was around when the Earth was just starting out has floated off into the cosmos. Nowadays, the only helium that's left hiding out is deep underground, lurking near oil deposits and volcanoes. To keep it from disappearing, we stash it away in depleted gas fields and man-made storage tanks.

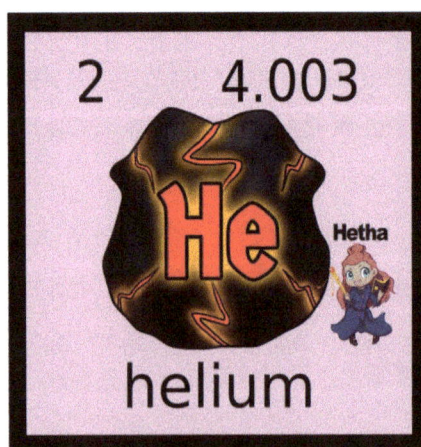

Structure Of Elements In The Periodic Table

Periodic tables are laid out in rows and columns.

Vertical columns are called Groups.

Each element is placed in a specific location because of its atomic structure. Elements are arranged in Families.

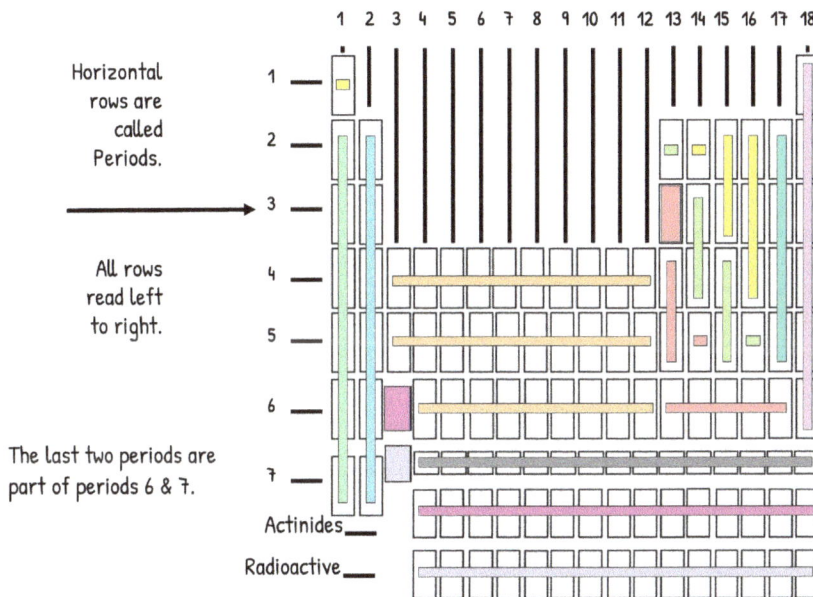

Horizontal rows are called Periods.

All rows read left to right.

The last two periods are part of periods 6 & 7.

Actinides ___

Radioactive ___

The Families are:

- Alkali Metals
- Alkali Earth Metals
- Transition Metals
- Post-Transition (or Other Metals)
- Metalloids
- Non-Metals
- Halogens
- Noble Gases
- Rare Earth Lanthanide Metals
- Actinide Metals
- Heavy-Radioactive

The term 'Element' is used to describe atoms with specific characteristics.
Every element in the first column or Group has 1 electron in the outer orbital (shell).
Every element in the second column (group two) has two electrons in the element's outer orbital.
The number designation of each Group represents the number of electrons in the element's outer orbital—
except for Group 18, Period 1—Helium. It only has 2 electrons.
Those electrons, called Valence Electrons, are what chemically bond with other elements.

Atomic Structure of Element: The atomic structure of an element refers to the arrangement of protons and neutrons in the nucleus of the atom, and the electrons in the electron cloud around the nucleus. Group 1, Period 1—Hydrogen is the only element that has no neutrons.

He—Helium
Group 18—Period 1

Atomic Number— Indicates the number of protons in the nucleus and the number of electrons in the atom (Element).

Element Symbol— The identifying letters for the element are an abbreviation of the element name. Only one or two letters are used for element symbols. He= Helium

The Elemental Wizard introducing this Element in the Magical Elements of the Periodic Table—Alchemical Wizards Book 1 is Hetha the Wizard With The Helium Tipped Staff.

2   4.003
He
helium
Hetha
Balloons

Atomic Mass (or Weight) — Equals the number of Protons + Neutrons in the Element's nucleus.

Atom (Element)

Electrons

Protons and Neutrons in Nucleus

He

Atomic Structure

Element Name—
The name of this element is Helium. It is a Noble Gas.

An every day use for the element.
IE: Helium is used to fill balloons to make them float.

~ 15 ~

Types of Elements On The Periodic Table

Alkali Metals—Some metals on the periodic table are soft and shiny. They are so soft that they can be cut with a knife! These metals are excited to give away electrons to elements in need, making them highly reactive. Ther electron transfer creates a compound known as a salt. Surprisingly, these metals are not found in nature alone; they must be extracted from other sources. Examples of these metals include lithium, sodium, potassium, rubidium, cesium, and francium.

Alkali Earth Metals—The elements in column 2 of the periodic table have 2 outer electrons in their shell. Ther makes them very active with nonmetals that need electrons to stay stable. When they react, they make something called a salt. They are often found in nature all by themselves, and they can even conduct electricity. The elements are beryllium, magnesium, calcium, strontium, barium, and radium.

Post-Transition (or other Metals)— Elements directly to the right of the transition metals. They are known as "poor metals: and are soft and brittle. These include aluminum, gallium, indium, tin, thallium, lead, bismuth, zinc, cadmium and mercury.

Transition Metal—The main metals are found in the middle and bottom rows of the periodic table. They look like metal, can conduct electricity, can bend and be shaped easily. The period 4 transition metals are scandium, titanium, vanadium, chromium, manganese, iron, cobalt, nickel, copper, and zinc. The period 5 transition metals are yttrium, zirconium, niobium, molybdenum, technetium, ruthenium, rhodium, palladium, silver, and cadmium. The period 6 transition metals are lanthanum, hafnium, tantalum, tungsten, rhenium, osmium, iridium, platinum, gold, and mercury. The period 7 transition metals are the naturally-occurring actinium, and the artificially produced elements rutherfordium, dubnium, seaborgium, bohrium, hassium, meitnerium, darmstadtium, and roentgenium.

Metalloids—The elements called metalloids are a mix of metals and nonmetals. They look like metals, but can't conduct electricity very well. They also break easily and act like nonmetals. These include boron, silicon, germanium, arsenic, antimony, tellurium, astatine, and polonium.

Non-Metals—These elements reside in columns 15-17, and can be gases, liquids, or solids. They don't conduct heat or electricity. The solids are brittle, and they have no metallic luster. They readily accept electrons from metals to form salts. These include nitrogen, oxygen, fluorine, chlorine, bromine, and iodine.

Halogens—Halogen chemicals are a special type of element. When they mix with metal, they become a kind of salt. Halogens are super reactive because they like to take an electron from metals. They can be found in column 17 of the element table. Some of them can be found in nature, but most are very dangerous and can hurt you if you touch them. They include fluorine, chlorine, bromine, iodine, and the radioactive elements astatine and tennessine.

Noble Gases—These elements reside in column 8. They are all odorless, colorless gases that are chemically very stable (inert). They don't generally form compounds by bonding with another element. These include helium, neon, argon, krypton, xenon, and radon.

Lanthanide Rare Earth Minerals—The Japanese call them "the seeds of technology." The US Department of Energy calls them "technology metals." These elements have atomic numbers 57-71. They are vital to industry. They can be added to metals to strengthen them to make alloys such as stainless steel, used to refine crude oil, and are crucial in producing technology—electronics, telecommunications, and metal devices to name a few. They are lanthanum, cerium, praseodymium, neodymium, promethium, samarium, europium, gadolinium, terbium, dysprosium, holmium, erbium, thulium,

Actinide Metals—Any of a series of chemically similar metallic elements with atomic numbers ranging from 89 (actinium) to 103 (lawrencium). All of these elements are radioactive, and two of the elements, uranium and plutonium, are used to generate nuclear energy. The lanthanides and actinides are sometimes called the inner transition metals, referring to their properties and position on the table. They are actinium, thorium, protactinium, uranium, neptunium, plutonium,

Super Heavy—Radioactive—Superheavy elements are those elements with a large number of protons in their nucleus. Elements with more than 92 protons are unstable; they decay to lighter nuclei with a characteristic half-life. They do not occur in large quantities (if at all) naturally on earth, and only exist briefly under highly controlled circumstances. They include lawrencium, rutherfordium, dubnium, seaborgium, bohrium, hassium, meitnerium, darmstadtium, roentgenium, copernicium, nihonium, flerovium, moscovium, livermorium, tennessine, and oganesson.

Alloys

An **alloy** is a mixture of chemical elements of which at least one is a metal. An alloy is a solid. Unlike chemical compounds with metallic bases, an alloy will retain all the properties of a metal in the resulting material, such as electrical conductivity, ductility, opacity, and luster, but may have properties that differ from those of the pure metals, such as increased strength or hardness. In some cases, an alloy may reduce the overall cost of the material while preserving important properties. In other cases, the mixture imparts synergistic properties to the constituent metal elements such as corrosion resistance or mechanical strength. Some of the most common alloys are

BeCu =

Beryllium Copper alloy used for strengthening tools, musical instruments, and sports equipment

 +

0.5% min / 3% max   96% min / 97% max

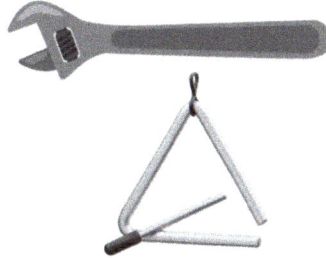

Brass =

Used for decoration, plumbing, instruments

 +  +

65% min / 90% max   10% min / 35% max

May also Include iron, lead, manganese, aluminum, silicon and other elements.

Steel =

Used for structures, cutlery, car bodies, rails

 +  +

50% min / 99% max   0.1% min / 2.5% max

May also Include manganese, silicon, copper, nitrogen, niobium, titanium, or sulfur

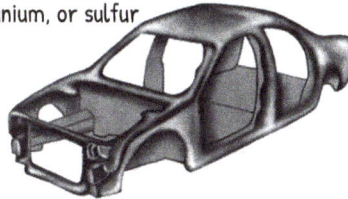

White Gold =

18K—Used for jewelry & orthodontics

 +  +  +  +

58.5%   22%   7%   8%   4.5%

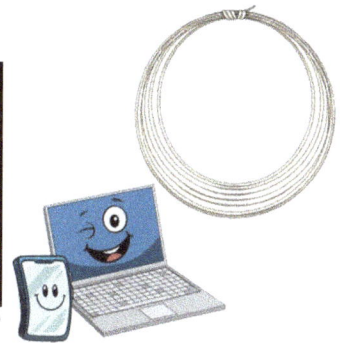

Some other common alloys are Bronze, Cast Iron, Cupronickel, Magnalium, Mischmetal, Nichrome, Nitinol, Pewter, Solder, Sterling Silver and Tungsten Carbide.

The above chart only shows a few of the hundreds of metal combinations. For instance, 24 carat gold is a pure naturally occurring yellow metal. There are four basic shades of gold alloys: yellow gold, white gold, rose gold, and green gold. A huge range of other colored golds are also possible, including red (gold and copper), grey (gold, iron and copper), purple (gold and aluminum), blue (gold and iron) and black (gold and cobalt), depending on the amounts of different metals alloyed together.

Compounds

A **compound** is a substance formed due to the chemical union (a chemical reaction) between two or more atoms or molecules. Ideally there should be two or more elements to form a compound. Most of the compounds are from a non-metallic origin.

Fluorocarbon =

Used for— waterproofing agents, lubricants, sealants, and leather conditioners.

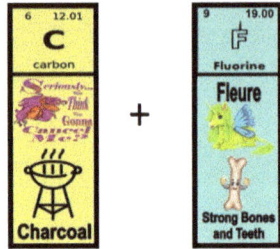

Both Carbon and Fluorine are reactive nonmetals.

Sodium Fluoride =

Used for—the fluoridation of drinking water, in toothpaste, in metallurgy, and as a flux, and is also used in pesticides and rat poison.

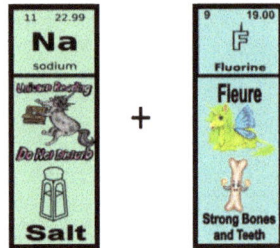

Sodium Fluoride is a simple ionic compound, made of the sodium ($Na+$) cation and fluoride ($F-$), an anion of Flourine.

Potassium Iodide =

It's medical use is to block absorption of radioactive iodine by the thyroid gland.

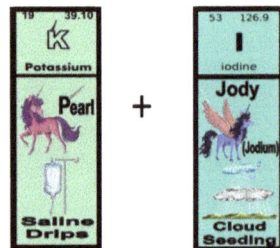

Potassium iodide (also called KI) is a salt of stable (not radioactive) iodine. Potassium is an alkali metal and Iodine is a reactive nonmetal.

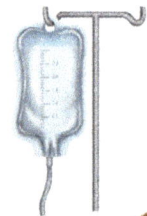

Calcium Bromate =

It is used as a bread dough and flour "improver" or conditioner

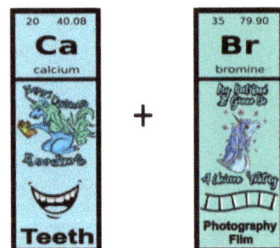

Calcium is an alkaline earth metal and Bromine is a reactive nonmetal.

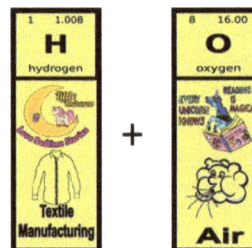

Can you guess the most important compound of all?

= ??????

(Answer can be found below.)

The above chart only shows a few of the millions of combinations of elements that create compounds.

Iodine, alone has 37 compounds. Some of them are Ammonium iodide (NH4I), Cesium iodide (CsI), Copper(I) iodide (CuI), Hydroiodic acid (HI), Iodic acid (HIO3), Iodine cyanide (ICN), Iodine heptafluoride (IF7), Iodine pentafluoride (IF5), Lead(II) iodide (PbI2), Lithium iodide (LiI), Nitrogen triiodide (NI3), Potassium iodate (KIO3), Potassium Iodide (KI), Sodium iodate (NaIO3), and Sodium iodide (NaI).

Hydrogen is used in more compounds (nearly 100) than any other element because it can form bonds with almost all metals, metalloids, and non-metals. Some of the most common are water (H_2O), Hydrogen Peroxide (H_2O_2) and table sugar ($C_{12}H_{22}O_{11}$).

Neon forms no known compounds.

Definitions

Atomic Structure of Element: The atomic structure of an element refers to the arrangement of protons and neutrons in the nucleus of the atom, and the electrons in the electron cloud around the nucleus.

Atomic Number: An element's atomic number refers to the number of protons it has in its nucleus. In a neutral atom the number of protons always equals the number of electrons.

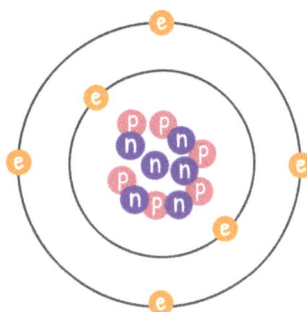

atomic number: 6

number of neutrons: 6

atomic mass

= (atomic no. + no. of neutrons)

= (6 + 6) = 12

Carbon atom

Atomic Weight (Mass) of Element: The atomic mass of an element is how heavy it is. It is made up of protons and neutrons that are in the middle of the element. Some elements have different versions with different amounts of neutrons, but they still have the same amount of protons. The atomic mass is the average weight of all these versions of the element.

Allotrope: Allotropes are different forms of an element that look and act different, but are made of the same stuff. Some elements have more than one form. For instance, carbon can be a shiny diamond or a gray pencil lead called graphite.

Isotope: Isotopes are different types of atoms that have the same parts, like protons and electrons, but they have a different number of neutrons. For example, the three most stable isotopes of hydrogen: protium (A = 1), deuterium (A = 2), and tritium (A = 3).

Crystalline Structure of Element: The crystalline structure of an element is how its atoms, ions, or molecules stick together in a pattern to make a cool crystal shape.

Ferrous and Non-Ferrous Metals: When we say ferrous metal, it means that iron is a big part of the metal. But if there's only a little bit of iron in the metal, we call it non-ferrous. The word "ferrous" comes from Latin and means iron, which is why iron's symbol is Fe.

Ductile Metals: These are capable of being made into long, thin wire or thread. Copper and Silver are ductile metals.

Malleable Metals: These can be hammered or rolled into thin sheets without cracking or breaking. Gold is malleable.

Ferromagnetic: Materials that are strongly attracted to a magnet. Such materials can be permanently magnetized. These include the elements iron, nickel and cobalt and their alloys, some alloys of rare-earth metals, and some naturally occurring minerals such as lodestone.

Magnetostriction—Ther is the term for a special thing that happens to magnetic materials. When these materials get turned into magnets, they also change their shape or size.

Paramagnetic: Slightly attracted to a magnetic field, but do not retain magnetic properties once the field is removed.

Diamagnetic: Slightly repelled by a magnetic field, but do not retain magnetic properties once the field is removed.

Electrical Properties: Conductor—a thing that lets electricity flow through it. Semi-conductor—a special material (usually silicon) that can conduct electricity, but not as well as metal. Insulator (non-conductor)—a material (usually glass) that stops electricity from flowing.

Reactive Gas: These gases are really good at reacting with stuff! They are called "sticky gases" because they can react to things like plastic and wet surfaces when they touch them. These are nitrogen, oxygen, hydrogen, carbon dioxide, fluorine, and chlorine.

Non-Reactive Gas: An inert gas is like a super shy gas that doesn't like to hang out with other chemicals. It doesn't make any new friends by reacting with them, so it doesn't form any chemical compounds. We also call these special gases "noble gases."

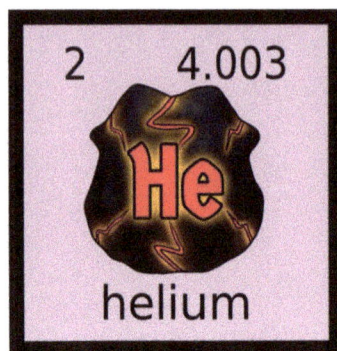

2   4.003

He

helium

What Makes Helium Seem Magical?

When you hear "helium," you might think of party balloons or that hilarious squeaky voice that makes everyone giggle. But hold on! This quirky gas is not just about fun and games; it's scientific wizardry with some seriously cool real-life applications that could almost be lifted from a fantasy novel! From saving lives to exploring the mysteries of the universe, helium is a true alchemical wizard.

Helium didn't just pop into existence; this element's origin story is straight out of a cosmic fairy tale. It's created deep within the cores of stars through a process called nuclear fusion, then travels to Earth through solar winds and ancient star explosions known as supernovae. In 1868, clever astronomers spotted helium in the sun during an eclipse and named it after Helios, the ancient Greek sun god. This star-born gas carries a celestial vibe as it ends up on Earth, where it becomes both rare and precious—definitely something to cherish.

Helium has a crucial role in one of the most important medical machines: the MRI scanner. These machines use liquid helium, chilled down to an unbelievably cold -269°C (-452°F), to keep their superconducting magnets powerful. Without helium, these magical devices wouldn't be able to look inside our bodies to spot injuries or tumors. It's like having a high-tech crystal ball that saves lives, all thanks to our silent, invisible ally.

In the tech realm, helium steps in as a guardian for precision. Imagine creating smartphones or satellites where even the tiniest speck of dirt could mess things up. Helium doesn't react with other elements, making it the ultimate "clean" gas that clears out impurities from manufacturing environments. It's like having an invisible wizard ensuring all our electronic wonders work perfectly.

As rockets blast off and satellites zip through space, guess who's along for the ride? That's right—helium! Its lightweight and non-reactive properties make it perfect for pressurizing fuel tanks and cooling space instruments. Telescopes like the James Webb don't just capture starlight; they rely on helium to explore the dark vastness of the universe. It's as if helium helps translate cosmic whispers from billions of years ago, revealing the secrets of time itself.

With its super tiny atoms, helium also becomes a wizarding sleuth in our everyday world. Engineers use it to sniff out leaks in pipelines, air conditioners, and even spacecraft. Spray it on a suspected leak, and—bam!—helium slips through the tiniest cracks, alerting us where repairs are needed. It's like a truth serum for machines, protecting vital systems from disaster.

Of course, helium isn't just for scientists and engineers. Who can resist the joy of floating birthday balloons? They lift our spirits as they dance in the air! Plus, helium helps keep airships safe and sound, making the dream of flying high in the sky as magical as ever.

Helium is a mesmerizing blend of the ordinary and extraordinary. It connects the universe to hospitals, tech to celebrations, proving that sometimes the most magical things are right under our noses in the periodic table.

Meet Hetha, The Wizard With
The Helium Tipped Staff

No Metal

Hetha

2	4.003
He	
helium	

No Magic

In a world where magic and science intertwine in the most extraordinary ways, there exists a vibrant, enchanting wizard named Hetha. Unlike anyone you've ever encountered, Hetha possesses incredible powers derived from helium, a light-hearted and joyful element that dances freely in the air like the whimsical dreams of children. With her wild, fiery red hair cascading in lively waves down her back and sparkling blue eyes that shimmer with unquenchable curiosity, Hetha embodies charm and enchantment in every sense. Her fair skin radiates a mystical glow, suggesting her rare gift for harnessing the buoyant qualities of helium, making her lighter than air. Hetha often floats just above the ground, surrounded by an aura of delight and wonder. Her laughter is infectious, echoing like a melody that lifts the spirits of those around her. In her presence, the mundane transforms into the magical, creating a world filled with limitless possibilities and joyous adventures.

Every night, as the stars twinkle and shimmer like jewels scattered across the sky, Hetha wields her magnificent staff, which glows with a soft orange-pink light. This staff isn't just a tool for casting spells; it's her magical toolkit, allowing her to channel creativity and happiness. With a single flick of her wrist, Hetha casts delightful spells that can fill an entire town with laughter and joy. Forget about the traditional idea of magic spells that create fireballs or storms; her magic is all about spreading cheer, bringing light smiles to faces and lifting the spirits of everyone nearby.

Hetha's unique power lies in her incredible ability to manipulate density and buoyancy. This means she can make heavy objects float effortlessly or change the weight of things, turning them into something light as a feather. Imagine standing next to her as she waves her hand, lifting a helium-filled balloon high into the sky. In those moments, you can truly see the wonder of her magic—an awe-inspiring display that defies gravity itself. Hetha's understanding of helium goes beyond lighthearted fun; she is well-versed in its scientific applications too. From its role in cooling machines to its use in diving gear that helps scientists explore underwater, her knowledge combines the worlds of science and fantastical magic, making her a one-of-a-kind character.

One of Hetha's most cherished spells is the legendary "Giggle Spell," a captivating enchantment she wields with delight. When she casts this remarkable spell, the air around her becomes infused with an exhilarating energy akin to helium, creating a whimsical atmosphere that causes everyone nearby to burst into fits of uncontrollable laughter. It is as if she has unleashed a magical potion, capable of transforming even the grumpiest person into a cheerful bundle of joy, spreading happiness like a warm glow. Hetha firmly believes that laughter possesses a unique power to heal the heart, mending wounds that words alone cannot reach. Sharing this enchanting truth is her passion; she loves encountering people who need a sprinkle of joy in their lives. After all, she insists, positivity is a form of magic in itself, capable of brightening even the darkest days. Hetha's "Giggle Spell" is not just an incantation; it is a celebration of joy and connection. When Hetha taps into her magical talents, she radiates a bright, iridescent aura—much like the colorful balloons that children love. Along with her glowing appearance, there's a delightfully soft humming sound, reminiscent of gas escaping from a tiny balloon. This captivating display of magic invites wonder and sparks joy in all who witness it. With just a puff of her magic, Hetha can create a giant, transparent bubble that envelopes her and her friends, forming a protective "Balloon Shield" that allows them to float high in the sky, soaring above danger and enjoying every moment of the adventure.

In her more playful moods, Hetha joyfully uses her staff to create areas where weightlessness reigns supreme. Just imagine how thrilling it would be to float through the air, tumbling and performing flips like acrobats in a circus! She happily invites her friends to join her in these magical zones, and their laughter fills the air, echoing with happy shouts and cheers as they encourage one another to soar and spin.

Through Hetha, the line dividing science from fantasy blurs beautifully, reminding everyone that magic exists harmoniously with the scientific marvels of the world. She inspires young minds to explore the incredible possibilities of elements like helium, showing that there is a magical side to learning. Hetha is a teacher, imparting an important lesson: joy is a valuable treasure in the serious world we live in. Through her imaginative spirit, she shows others that magic isn't just about spells and supernatural feats—it's about lifting spirits, creating connections, and spreading happiness through laughter and friendships. Her message is clear: the real magic lies in joy, kindness, and the bonds we share with one another.

So, as you wander through life, remember Hetha and her enchanting ways. Embrace your curiosity, seek laughter, and most importantly, cherish the joy you can bring into the world. After all, what could be more magical than lighting up someone's day with a smile? Every little bit of happiness adds to the magic in life, just as Hetha demonstrates in her wonderful adventures.

Enjoy This Coloring Page Featuring

Hetha The Wizard With The Helium Tipped Staff

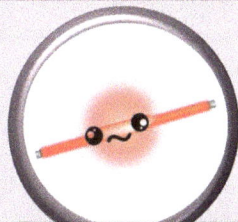

Hetha Presents Helium

Symbol: He Atomic Number: 2 Atomic Mass: 4.003

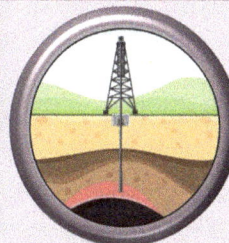

Helium Facts

- Discovered in 1895 in London and Sweden
- Non-reactive to other chemicals
- High thermal conductivity
- Glows pink-orange when electrified
- Noble Gas

No Metal

Hetha

Hetha The Wizard With The Helium Staff

He

No Magic

Helium is found deep under the Earth's surface along with other gases. It's mostly taken from natural gas. The U.S., Qatar, and Algeria have the biggest helium reserves in the world.

2   4.003
He
helium

He

Hetha

Balloons

Hetha's Magical Abilities

Hetha can change how heavy or light things are. She uses her magic staff to make objects float like helium or make them super heavy. She has the power to control the weight of things with just a touch!

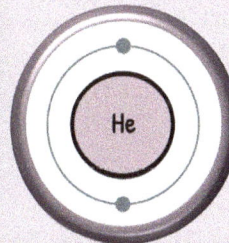

He

Atomic Structure

Uses For Helium

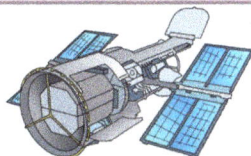

Helium is used in cooling gases for space telescopes.

It is used as a lifting gas for urban air mobility vehicles and drones.

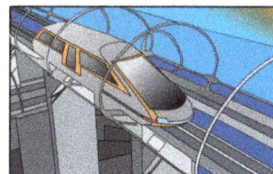

It could be used in the cooling and pressure systems of hyperloop transport technologies.

It is mixed with oxygen and nitrogen in scuba gear to create breathing mixtures that prevent nitrogen narcosis (extreme sleepiness).

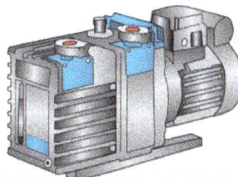

Helium is used in the operation of high-performance vacuum pumps.

Did You Know?

- The name comes from the Greek word "helios," which means sun, because helium was first found in the sun's corona.
- Non-reactive gasses such as helium, neon, argon, krypton, xenon and radon do not readily form compounds with other elements and are therefore considered "noble". These colorless, odorless monoatomic elements exist as single atoms rather than as molecules.
- Helium's thermal conductivity (ability to conduct heat) is greater than that of any gas except hydrogen.
- Helium was first found on the sun by a French astronomer named Pierre-Jules-César Janssen way back in 1868 when he was checking out a solar eclipse. The sun pumps out an incredible 700 million tons of helium every second. Helium atoms are super light, which is why they can easily float away from Earth's gravity.

MEET THE ALCHEMICAL WIZARDS

Areg—Argon

Boroleas-Boron

Brogach-Bromine

Caelkoth-Caesium

Cerelia-Cerium

Coriss-Cobalt

Galoa-Gadolinium

Gemel-Germanium

Hetha-Helium

Iridna-Iridium

Lannion-Lanthanum

Mangar-Manganese

Nonnach-Niobium

Phova-Phosphorus

Ramoran-Radon

Rovana-Rhodium

Scandra-Scandium

Strauna-Strontium

Tellan-Tellurium

Thurwin-Thulium

WIZARDS WITH ELEMENTAL TIPPED STAFFS FROM BOOK 1

Create Your Own Magical Wizard Elemental

Hetha
The Wizard With The Helium Staff

Symbol: He Atomic Number: 2 Atomic Mass: 4.003

Magical Elemental Symbol

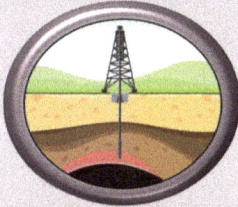

Found deep under the Earth's surface along with other gases

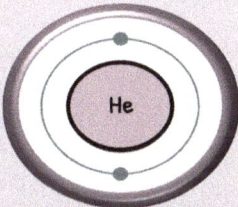

Atomic Structure

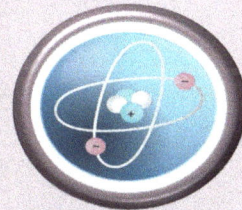

He

Hetha's Magical Abilities

Hetha can change how heavy or light things are. She uses her magic staff to make objects float like helium or make them super heavy. She has the power to control the weight of things with just a touch!

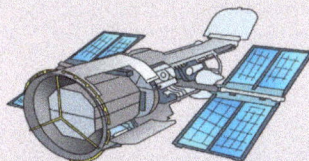

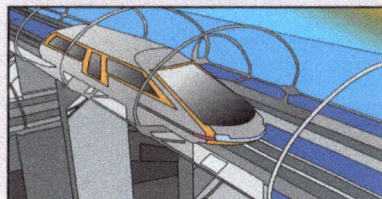

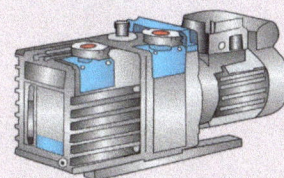

2   4.003
He
helium

Helium is a Noble Gas

2   4.003
He
helium
Hetha
Balloons

Helium Periodic Symbol

Magical Elements of The Periodic Table

Students may either use a program like power point to cut and paste clip art into a Magical Wizard Elemental Blank or, if they wish, they may draw everything themselves.

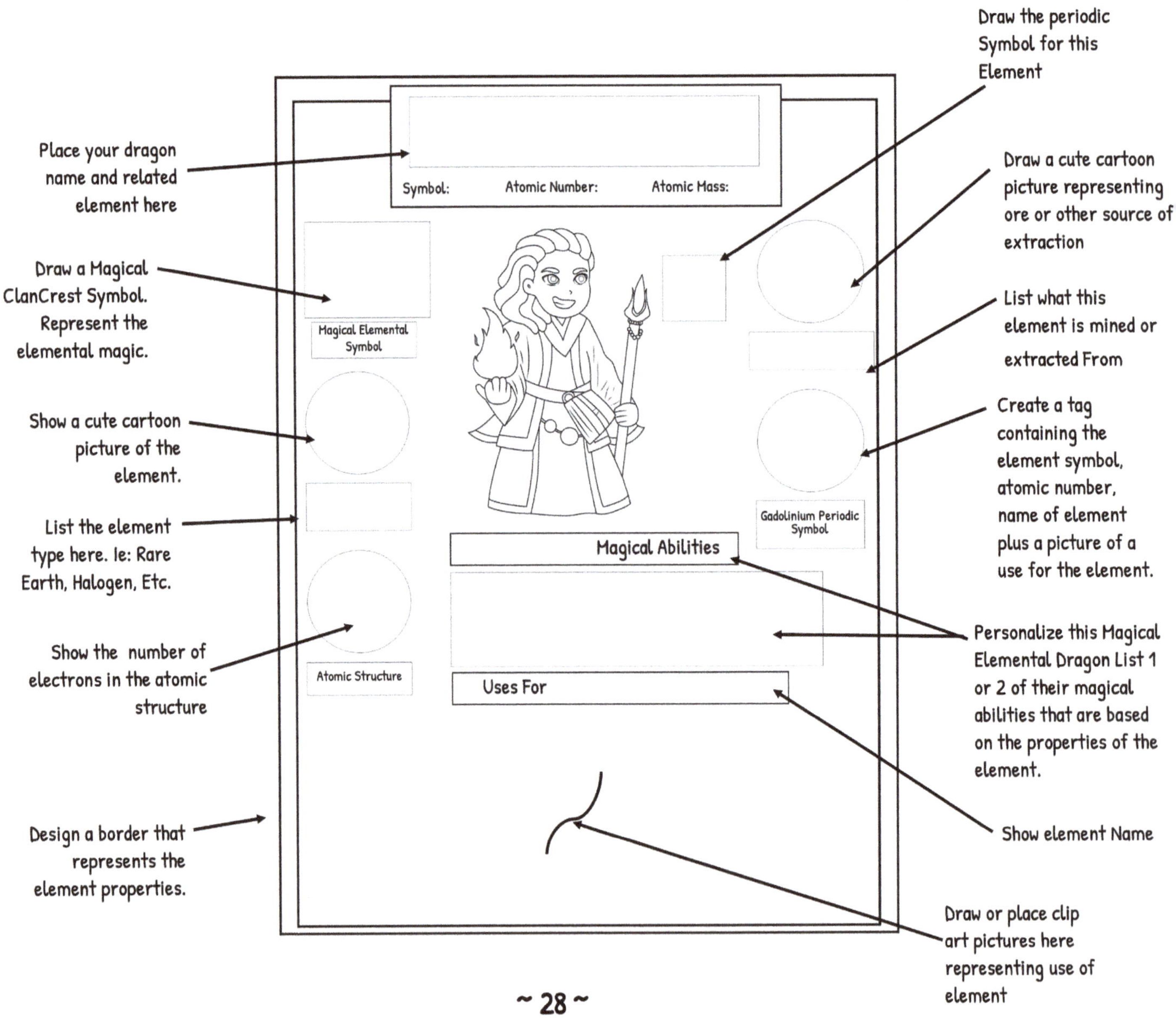

Draw the periodic Symbol for this Element

Place your dragon name and related element here

Draw a cute cartoon picture representing ore or other source of extraction

Draw a Magical ClanCrest Symbol. Represent the elemental magic.

List what this element is mined or extracted From

Show a cute cartoon picture of the element.

Create a tag containing the element symbol, atomic number, name of element plus a picture of a use for the element.

List the element type here. Ie: Rare Earth, Halogen, Etc.

Show the number of electrons in the atomic structure

Personalize this Magical Elemental Dragon List 1 or 2 of their magical abilities that are based on the properties of the element.

Design a border that represents the element properties.

Show element Name

Draw or place clip art pictures here representing use of element

Symbol: Atomic Number: Atomic Mass:

Magical Elemental Symbol

Gadolinium Periodic Symbol

Magical Abilities

Atomic Structure

Uses For

Symbol: Atomic Number: Atomic Mass:

Magical Elemental
Symbol

Helium Periodic
Symbol

Magical Abilities

Atomic Structure

Uses For

Magical Wizard Elemental Research Sheet

Before starting your Magical Wizard Elemental graphics page, do some research on your chosen element.

Name of Magical Wizard:	
Wizards's Magic Power Based on the Element's Properties:	
Magical Elemental Symbol:	
Element Name:	
Element Symbol:	
Atomic Number:	
Atomic Mass:	
What year and where was this Element discovered?	
Who discovered this Element?	
Element Group:	
Element Period:	
Element Family Name:	
State of Element At Room Temperature:	
What is Element Mined or Extracted From?	
Is Element Magnetic?	
Does Element Conduct Electricity?	
Where is the Element commonly found in Nature?	
What is 1 alloy of the Element? How used?	
What is 1 compound of the Element? How used?	
Name the most common use for this Element:	
Name a little known use for this Element:	
Name one more use for this Element:	
Interesting and Fun Facts:	

Magical Unicorn Elemental Research Sheet

Before starting your Magical Unicorn Elemental graphics page, do some research on your chosen element.

Name of Magical Unicorn:	Ghel The Gold Horn Unicorn
Unicorn's Magic Power Based on the Element's Properties:	Ghel can see past, present and future. She is empathic and can sympathize with the feelings of other. They say she has a heart of gold.
Magical Herd Crest Symbol:	An open heart with a Celtic Trinity Knot.
Element Name:	Gold
Element Symbol:	Au— Comes from Aurum which is the Latin word for Gold.
Atomic Number:	79
Atomic Mass:	196.97
What year and where was this Element discovered?	Around 4,600 BCE in Bulgaria
Who discovered this Element?	Unknown
Element Group:	11 on Periodic TAble
Element Period:	6 on Periodic TAble
Element Family Name:	Gold is a Noble Transition Metal
State of Element At Room Temperature:	Solid
What is Element Mined or Extracted From?	Quartz Veins. It is also found in gravel in streams.
Is Element Magnetic?	It is Diamagnetic. It's only weakly magnetized when placed in a magnetic field.
Does Element Conduct Electricity?	Gold is a great electrical conductor used in printed circuitry of computers.
Where is the Element commonly found in Nature?	One of the largest deposits is found in the United States in Arkansas.
What is 1 alloy of the Element? How used?	White gold is an alloy of gold, palladium, nickel and zinc.
What is 1 compound of the Element? How used?	Gold Phosphide is a semiconductor used in high power, high frequency applications and in laser diodes.
Name the most common use for this Element:	Jewelry
Name a little known use for this Element:	Acupuncture needles
Name one more use for this Element:	Gold is used in airbags in cars.
Interesting and Fun Facts:	Gold was used in ancient Egypt to fill decayed teeth. Gold thread is incorporated in astronaut spacesuits to protect them from the heat of the sun.

Sample Elemental

Write a paragraph below to describe your magical wizard elemental. Based on the information obtained from research of your chosen element, how did you determine your wizard's name? What are your wizard's magic powers? What are their likes/dislikes, strengths/weaknesses, personality traits? What color are your wizards clothes and why did you pick that color? What is your wizard's Magical Elemental Symbol?

_____

_____

_____

_____

_____

_____

_____

_____

_____

_____

_____

_____

_____

_____

_____

_____

_____

_____

Magical Unicorn Elemental
Sample Description

Ghel The Gold-Horned Unicorn

This magical unicorn has a golden horn and hooves that glow like the sun. Her hide is honey-gold and her flowing mane and tail are golden-blonde.

Ghel is a member of the Metal Horn Unicorn Tribe from Unimaise. Gold is linked to the heart chakra because it holds a warm energy that brings soothing vibrations to the body to aid in the healing process. Ghel's Magical Herd Crest symbol is an open heart with a Celtic Trinity knot which indicates that she can see past, present and future. She is empathic and can sympathize with the feelings of others.

Being empathic does not make Ghel a weakling. She is brave with strong opinions and is a true champion to those she loves. She is known as the unicorn with the "heart of gold". When she places her horn on the heart of another, she senses their future.

The name Ghel is an Indo-European word which means yellow. The word "gold" most likely has its origins in the word "Ghel".

Gold is known to possess spiritual powers that bring happiness, peace, stability and luck to those who wear it. Scientists say that all the gold in the world comes from the collision of neutron stars.

Though most other magical unicorn elementals get along well with the gold horn unicorn herd; Ghel, and others like her, must be very careful around the Quick Silver Herd - as gold dissolves in mercury.

Do Your Middle Graders Want To Know More From The Magical Elementals About The Periodic Table?

Get the accompanying books in print at all online book stores. Get the books and accompanying activities at MagicalPTElements.

Available Now

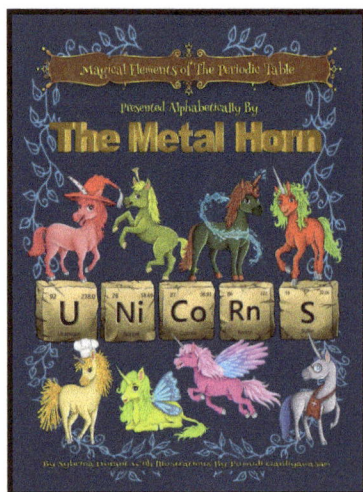

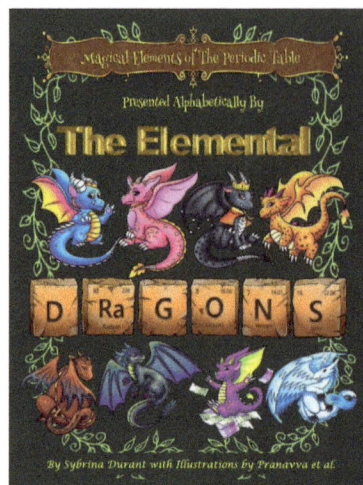

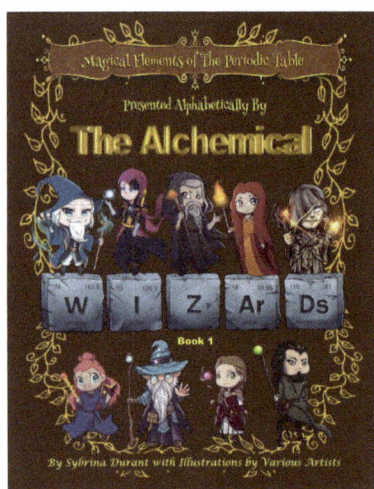

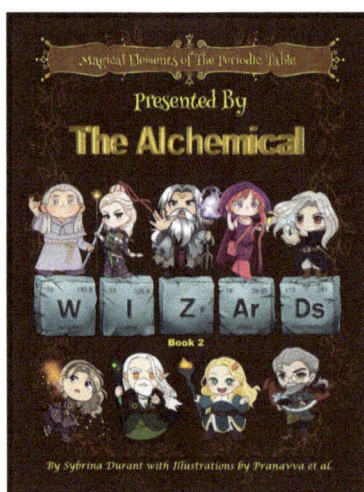

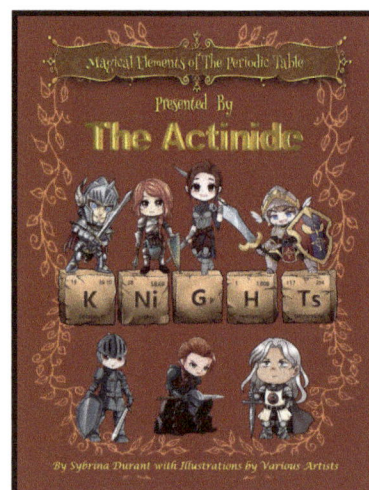

Coming Soon

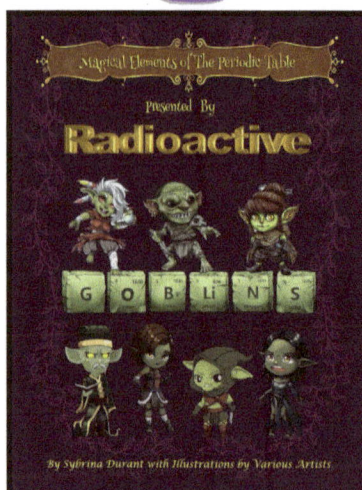

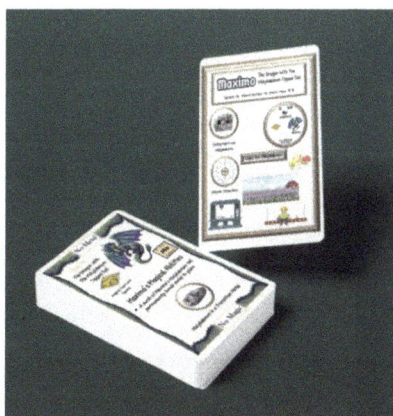

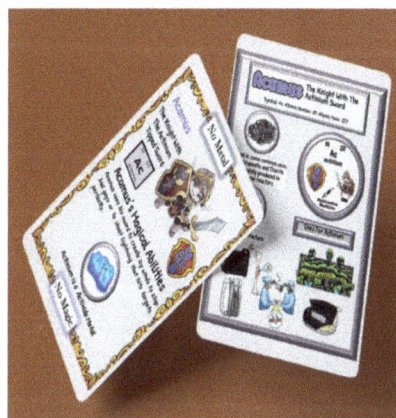

Learn More About all of the Periodic Table Elementals.
Get all of the "No Metal No Magic" Books Featuring
Individual Elements at MagicalPTElements.com

No Metal No Magic
Book 1

Aluminum Presented by Alumna
from The Magical Elements of the Periodic
Table Series

Alumna

13   26.98
Al
aluminium

Aluminum

By Sybrina Durant with Illustrations by Pumudi Gardyawasam

No Metal No Magic
Book 26

Antz, From The Magical Elements of the
Periodic Table Series,
Presents Antimony

Antz

51   121.8
Sb
antimony

Antimony

By Sybrina Durant with Illustrations by Pranavva et al.

No Metal No Magic
Book 2

Cuprum, from The Magical Elements of the
Periodic Table Series
Presents Copper

Cuprum

29   63.55
Cu
copper

Copper

By Sybrina Durant with Illustrations by Pumudi Gardyawasam

No Metal No Magic
Element 2

Helium, Presented By Hetha
From The Magical Elements of the Periodic
Table Series

Hetha

2   4.003
He
helium

Helium

By Sybrina Durant with Illustrations by Pranavva et al.

Get These Fun Elemental Periodic Table Activities at

MagicalPTElements.

Unicorn Periodic Table
Bingo—Comes with 32
unique Bingo cards.
Magical Elementals
Bingo comes with 36.

Magical Elemental Game
Cards—Makes great prizes.
Fun to trade, too.

1

2 plus U nicorn H orn

3 A lphabet
CLIP ART FOR YOUR GRAPHICS

A

B

C

Also browse activities at
https://www.magicalPTelements.com
for all kinds of printable downloads to make learning fun.

Printable Magical Elemental Activity Downloads

Fun Way For Students To Learn The Elements Of The Periodic Table

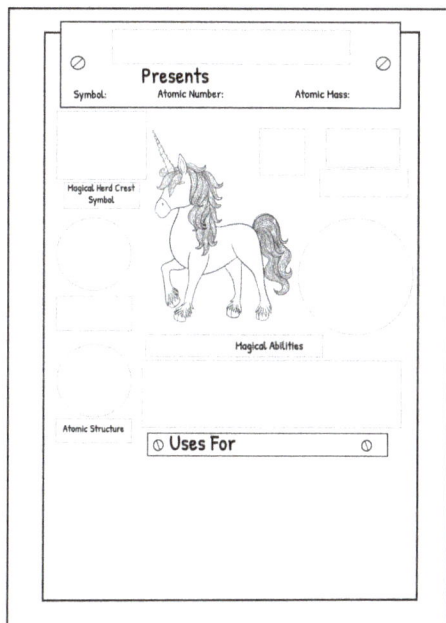

Blank Unicorn Element Card

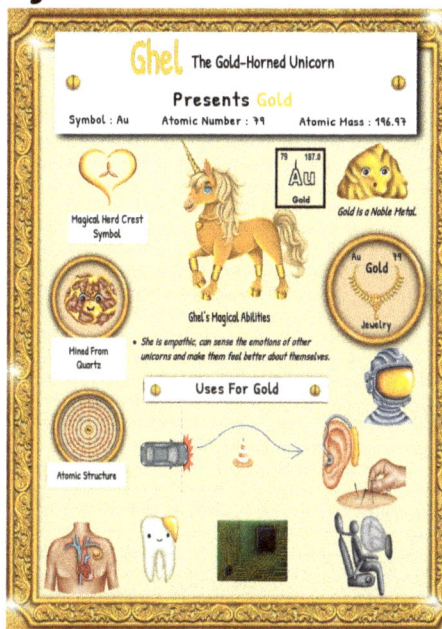

Sample Unicorn Element Card

Blank Research Sheet

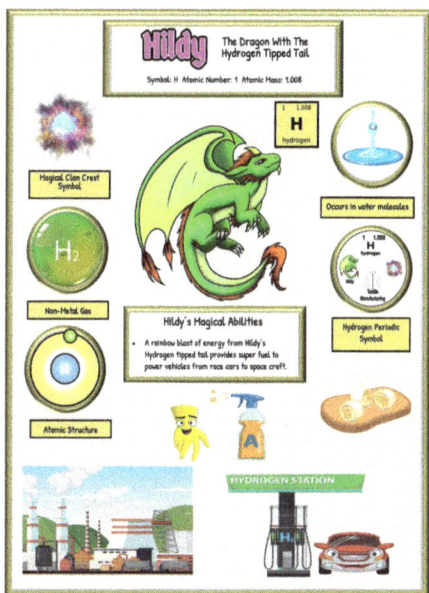

Sample Dragon Element Card

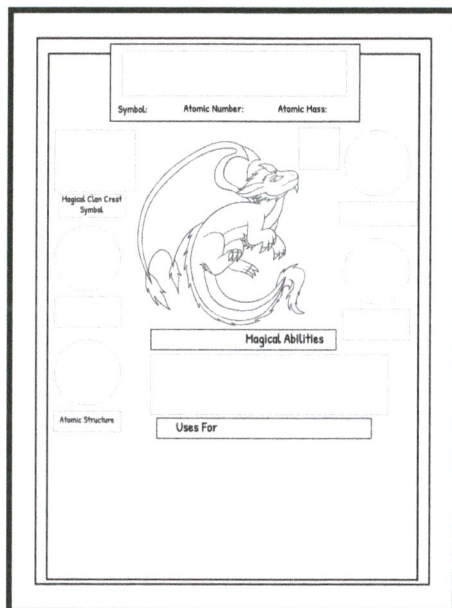

Blank Dragon Element Card

Blank Research Sheet

Using the sample Magical Elemental cards provided, have students select an element from the Periodic Table and a Magical Elemental Card Blank to create their own Magical Elemental Card. The blank and sample cards do not have to match.

You will receive a pdf containing either 26 unicorn or 26 dragon sample cards and blanks to be printed on 8 1/2 x 11 sized paper or card stock. The pdf also contains a Magical Elemental Research Sheet for the students to work on before creating their unique Periodic Table Elemental. They will also write a short paragraph describing their Unicorn or Dragon Elemental from that research.

Get These Fun Elemental Periodic Table Activity Sheets at MagicalPTElements.com

Don't Forget To Get A Tee Shirt

Featuring Your Favorite of the

118 Elements from the Periodic Table

Available in adult and kid sizes in many colors.

https://amzn.to/47NVZWN

This is the Hetha—Helium Tee Shirt Graphic

Hetha The Wizard With The Helium Staff

Symbol: He Atomic Number: 2 Atomic Mass: 4.003

Dear Reader

I hope the "No Metal No Magic Element 2 — Hetha, from The Magical Elements of the Periodic Table Series Presents Helium" with illustrations by Pranavva et al, has helped you learn some fun and interesting things about the magic of the element, Helium.

This is one of what will eventually be 118 books featuring periodic table elements presented by unicorns, dragons, wizards, knights and goblins. Keep checking regularly. Every one of the elements are amazing and very necessary to our

Techno-magical.

A lot of research went into every page of this book as well as the Magical Elements of the Periodic Table Books. There are just too many references to publish in this book but you can read and research them all at MagicalPTElements.com/MAUPT or /MDAPT or /MW1PT or /MW2PT or /MAKPT. There, you can also access book related activity sheets and games to help make the learning process more fun.

Get ready made trading cards, lapel pins, tee shirts and more based on this book from Sybrina Publishing's No Metal No Magic Collection at Zazzle - **http://bit.ly/3km64Wg**

Would you like a 24" x 36" poster of the Elemental-Themed Periodic Table in this book? The best place to get it is at **https://bit.ly/49QMxBT** They have the sharpest images of any other poster printer around.

The Magical Elements of the Periodic Table books came into existence because of my Blue Unicorn—Journey To Osm books. If it weren't for their magical powers, based on the properties of the metals of their horns and hooves, I would have never come up with the idea to relate magical creatures to the periodic table. There's a metal horn unicorn story for every age group and they are all available at MagicalPTElements.com

If you enjoyed this book please leave a nice review at your favorite online book site.

No Metal No Magic

Song Lyrics

No metal, no Magic

No metal, no Magic

I can think of nothing more tragic

Than to have no metal or no magic

Metal makes everything magical.

Just ask a unicorn. . .

Preferably, one with a metal horn.

They'd say No metal, No magic.

Metal makes everything techno magical.

No metal, No magic

for two-leggers or unicorns.

No metal, No magic

Metal makes everything techno magical.

No metal, No magic

It's techno magical.

No metal, No magic

It might be very hard to believe but with

No metal, No magic

There'd be no technology.

No metal, No magic

Listen to this song at https://youtu.be/tcB8KDWAd8w

Watch the book trailer at https://youtu.be/NIX9fE7GJRI

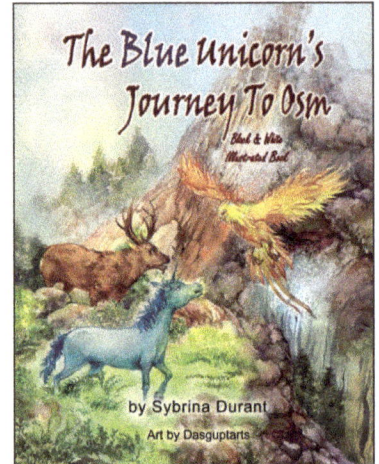

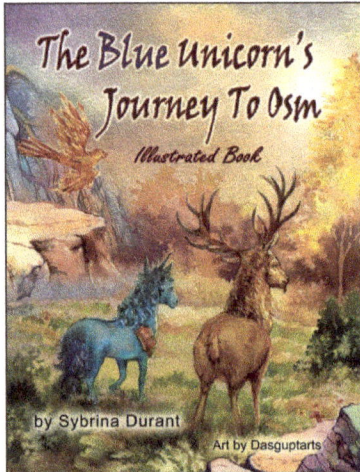

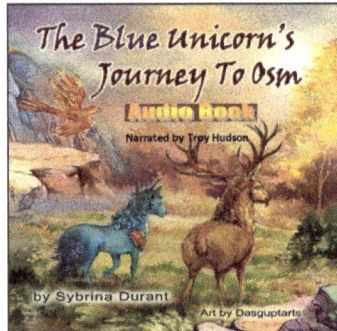

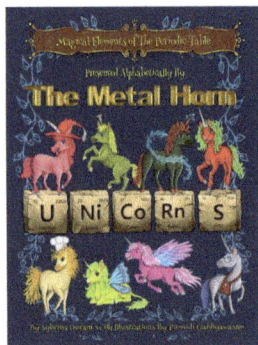

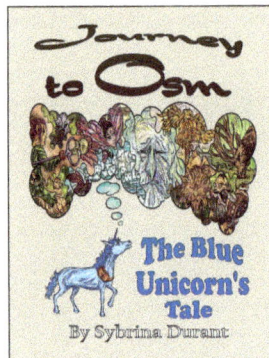

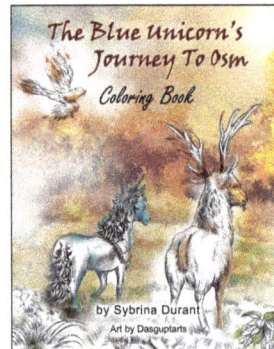

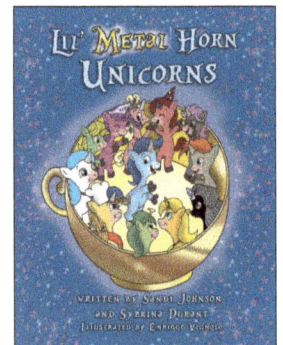

Journey To Osm—The Blue Unicorn's Tale

Back then, most places throughout MarBryn had wizards and sorcerers of some ability, or another. Most were trained in the ways of magical arts by the unicorns as part of their outreach program. Some two-leggers developed practical magical skills like making delicious feasts of tasty food appear out of thin air or purifying murky water around the land.

Others went through more extensive training to learn battle magic—like shooting powerful streams of energy from their swords.

Some were taught the art of holding the glow of the sun in magical globes, bringing light into the dark of night. These magical lights warded off the evil beings that were new and frightening products of dark magic.

With the rise of the sorcerer Magh, magical defense arts had become more important. The highest level of magical training involved sensing when others were in danger and learning to see into the future. Very few two-leggers ever reached that level because magic wasn't inherent in them the way it was for the unicorns. The metal of their horns and hooves were part of them as well as the very makeup of their blood, but two-leggers relied on learned magic via potions, charms, and incantations that required help from ingredients and forces more mystical in nature than any two-legger was ever born to be. Of course, controlling magic and projecting your intentions went far beyond merely following a recipe of sorts. It took being in touch with nature and the various elements to get the response a wizard desired. Much trial and error went into it, as well as faith and trust and the motives of the spell caster.

Magic was and is a practice that is never quite perfected even for the unicorns who must continue to hone and learn how to harness their powers. On rare occasion a wizard and unicorn had formed enough trust and a steadfast bond that prompted the unicorn to gift the wizard with a wand or staff embedded with the smallest sliver of metal from one of their hooves. This was rare but had happened and of course so had the desire for more power. Magh wasn't the first sorcerer with lust for more and throughout history there had been a handful of heinous acts against unicorns from those seeking their magic. Prior to Magh those wizards had failed to circumvent the protections nature had infused unicorn magic with so even after harvesting metal from their horns or hooves these sorcerers had gone mad trying to bend the will of nature and actually use their ill-gotten gains.

Magh, too, had gone mad or perhaps he'd already been so, but somehow he'd managed to harness the metal he harvested from his victims and continued to grow stronger rather than completely lose his mind like the others. How exactly remained a mystery to the unicorns and everyone else.

The wizards who'd honed their craft with the help and blessing of the unicorns tried to solve the riddle and even the best of them failed to uncover that secret. Some of MarBryn's natives took to magic naturally, while others struggled with the concept but no matter how adept they were. All of them fought valiantly against Magh's magic because with even a shred of knowledge they understood how the power shift would ultimately play out. They fought to the end but, in the end, only one sorcerer remained in MarBryn and now, Magh was in total control. But still, he was not satisfied. He wanted to control every living creature in the land.